Racecar Reecie

Elapsed Time

Kathleen L. Stone

ISBN–13: 978-1539444220
ISBN-10: 1539444228

Dedication
To my daughter-in-laws, Sheila and Devon. You have brought our family so much love and happiness!
I am so glad you are a part of our family!

Enjoy these other books by Kathleen L. Stone

Penguin Place Value
A Math Adventure

Number Line Fun
Solving Number Mysteries

Riley the Robot
An Input/Output Machine

Mason the Magician
Hundreds Chart Addition

Katelyn's Fair Share Picnic
More Math Fun

Money Tree Mysteries
Adventures with Quarters

Alien Even and Alien Odd
A Math Space Adventure

Kenley's Line Plot Graph
Another Math Adventure

Matthew's Sunshine Bakery
Multiplication Arrays

Firefighter Gary's Fire Safety Rules

Samantha's Search
3D Shapes

Grandma's Quilts
Fun with Fractions

Daniel's Day of Multiplication
Multiplication with Equal Groups

More Penguin Place Value
Hundreds, Tens, and One
s

Tick Tock Telling Time
Time to the Hour and Half Hour

Gavin the Gator
Greater Than and Less Than

Racecar Reecie
Elapsed Time

From My Quilted Heart to Yours
Heart Warming Quilts and Heart Healthy Recipes for Your Loved Ones

From My Quilted Heart to Yours Book 2
Quilts and Blocks from the Children's Book, Grandma's Quilts

Children, Fire, and Intervention
Creating a Program that Saves Lives and Communities

If you visit the Tucson racetrack
You will most likely see
A good friend of mine
Named Racecar Reecie.

Reecie absolutely loves to race!
Up and down hills he will climb.
He only has one goal in mind …
To do his best, every time!

On Sunday Reecie is racing
On a course down by the lake.
Let's see if we can find out
Just how long this race will take.

At *nine o'clock* he starts the race.
Reecie's really looking fine.
If it takes him *two hours*
When will he cross the finish line?

If you said *eleven o'clock*
You are exactly right!
Seeing that checkered flag at the end
Was really quite a sight.

Monday you will find Reecie
Beginning his race at *seven-thirty*.
It's a long and dusty course.
By the end he'll be pretty dirty.

It takes Reecie *three hours*
To finish Monday's race.
At *ten-thirty* he crosses the line
With a big smile on his face.

Tuesday Reecie is racing
In a field of wildflowers.
He begins the race at *noon*
And finishes in *four hours*.

Four hours past *noon.*
What time will that be?
Did you guess that *four o'clock*
Will be the time that we will see?

Wednesday's race was really short.
It only took a *half an hour*.
Just *thirty* short minutes
To race around the tower.

Reecie started at *twelve-thirty*.
He thought the race was fun.
What time do you think he finished?
You're right if you said at *one*!

At *one-thirty* on Thursday
Reecie raced around the block.
Thirty minutes later
He finished at *two o'clock*.

Friday's race was a tough one
With thunder, lightning, and rain showers.
His race began at *three o'clock*
And lasted for *five* hours.

From *three o'clock* to *eight o'clock*
Reecie raced through that storm.
But he sure felt proud
When he stood on the winner's platform.

Saturday is finally here.
For the last race of the week,
It will take Reecie *one* hour
And will end at a dried up creek.

He begins the race at *half past seven*
And finishes at *half past eight*.
After a whole week of racing
Reecie is sure looking great!

Tucson Racetrack

Home of
Racecar Reecie

So if you ever feel like racing
You might just want to spend
A little time with Racecar Reecie.
He's my very special friend.

Telling Time ... Elapsed Time

Elapsed time can be a tricky skill for children to learn but if you introduce it in steps and use visuals, children can master it easily. I like to begin using elapsed time in hour and half-hour increments. I also use a large "teacher clock" to demonstrate the hours as they go by. If the children have their own clocks to manipulate it will be even more effective.

Enrichment Activities

What Time Will It Be?

Materials needed:

Die or number cube
Wind-up clock (or other clock that child can move the hands on)
Index cards

Preparation of Materials:

Write a different time on each index card (be sure and write times to the hour and half-hour). If you are making your own number cube, you can write 1 hour, 2 hours, 30 minutes, etc. on the cube.

How to Play

Children turn over an index card, read the time, and set their clock to show that time. After rolling the die, they set their clock that many hours ahead. For example, if their clock is set at 2:00 and they roll a 6, they would set their clock to say 8:00.

Is It Time Yet

Materials needed:

Wind-up clock (or other clock that child can move the hands on)

How to Play

Tell children what time to set their clocks to. Share simple story problems for them to solve. For example, "The movie started at three o'clock and lasted for two hours. What time did the movie end?" (Children would set their clocks to 3:00 and then change the time to 5:00).

ABOUT THE AUTHOR

Kathleen Stone is a National Board Certified educator and is currently teaching second grade. *Racecar Reecie* is her seventeenth children's book. Born and raised in Washington State, she and her husband Gary live in the Olympia area. When not teaching, Kathleen can often be found quilting, sitting by the lake reading, or enjoying time with her family (especially her grandchildren)!

Math is all around us
No matter where you turn
Open your mind to the wonders of math
And all that you can learn